Math Pep Talk for College Students... who would rather quit or die first than take a math class

By

Susan Devine Napoli

To college students who struggle with math

Not associated with the American Mathematical Association or Math Awareness Month.

Contents

Introduction

I can't believe you would let a math class stop your dream.
Okay, more than one math class. Who's counting? You are. See,
you are doing math already. Your dream matters. It really does.
You are at a place in life where a decision like this changes
things. It matters a whole lot.

Come along and get yourself ready to take a math class. We
have some work to do before you enroll. You won't be doing
any math computations in this book, no story problems to work
and no formulas to learn. Just a long look at what you are
missing out on if you don't. Put down your paper and pencil. Put
away your calculator. Let's talk.

It is going to be okay. You dream is worth it.

 You can't, you say? You are one of those creative, spiritual,
intuitive folks? Yes we all have our preferences and interests.
Being right brained is not as bad thing as people might tell you.
You need these math tools for your dream. No dream can exist
without them, none. Even for us creative types. Everyone uses
their whole brain to function. Every. One. You can do this.

Part 1: Tell Your Math Story

"Ms. Napoli, I know that if you taught it, I would pass a math class." She said to me with pleading eyes. She meant it. Her name is gone with the passage of time but her message is with me tonight. I told her I wasn't qualified but the truth was I didn't want to let the demons out. You know the ones, those math teachers and experiences that come back to visit sometimes when I see numbers on a white page. I know how that goes for you, it is too hard to hear what the teacher is saying with them all talking in your head like that. I get that. I taught at a community college where we train people for careers and after all these years and completing my college degrees, I hear my math demons from time to time still.

Most teachers never could figure out how to help me. It started as a child when I moved from one state to another and I walked into something way more advanced than I had in first grade. Frustrated, some well-meaning second grade teacher asked me, "How come you can't do this?" What was a new second grader to say? So I told her what I knew, "We didn't finish the math book in first grade." They bought it. It became my mantra for years. I not only couldn't do the math but had to deal with public humiliation because of it too. I got called out in front of the class, dying at the chalk board again and again. I didn't know which was worse.

So they put me in a storage closet with the door closed and the light on to listen to the tape recording and write down the answers to math problems. It was hot in there. I spent most of the time wondering if they forgot me in there. As anyone who has any amount of common sense knows, you should never put a child in a closet with the door closed for any reason. It is harmful in lots of ways that have nothing to do with math.

I spent a large part of my life figuring out my way around it. Looking back it would have been easier to learn it than to spend that much energy in trying to avoid it. Only I didn't know that or have the support I needed.

There were some hopeful moments too. I sat in a class to learn how to teach math to elementary school children. This professor did something different. She brought in math manipulatives of all kinds. One day in particular she brought in geoboards and rubber bands. She had us put the rubber bands on in a certain way and then release them in a certain way. Then, I did mine and... pop. The area went into a rectangle where I could easily see what the area was. A light went on in a new room in my mind.

For years after that I had hoped to find another teacher like this who could help me to understand math this way. I never found one. I even went to one of those chain math tutoring places that had advertised they used math manipulatives. I walked in, paid, and they put a white page with numbers on it in front of me. Suddenly I was a little girl again not able to do math, feeling humiliated. I asked about the manipulatives. They pointed to the cabinet. I looked around the room. Everybody had white paper with numbers on it. I came back for the report but didn't continue.

For my Master's degree in Early Childhood Education I had to take a statistics class. That class was so hard for me. I worked and worked on that class. About 4 weeks from the end of the course, I suddenly figured out that math is a new language for me and I finally got that. I was fascinated by a video I saw that talked about statistics and how they used it to make potato chips. I saw that they take some potatoes from a truck that pulls up to the factory. They take some out and fry them up. In order to tell which ones they are, they punch a hole through the potatoes before slicing and frying them. When the chips come out done there is the hole and it is easy to pick them out because it has a browned edge. Then they taste them. If they taste good they buy the whole truck load. I finally understood what a sample size was. To this day, because of that video, I look for and remember to look for the potato chips with the hole in them when I buy chips. Sometimes they miss a few of these sample chips and they get bagged and sold. I have found several over the years. It delights me each and every time I find one because I know what happened and the math that goes with the hole in the chip. Math became meaningful in a new way for me that semester.

So what do I tell you this for? So that you can tell your math story and get it off your chest. Find someone you trust and get it all out. Or you could write it down on paper if there is no one like that for you. You have to get it out before we can begin. We are going to go on an adventure you and I. You have to lighten your load by setting that stuff down so there is even room in your head for the numbers. Don't worry, you won't need any paper and pencil at all in this book. That is the <u>last</u> step as I see it. We have a lot to do before you go sit down in a math class and they put white paper with numbers on it in front of you.

As a child care provider early in my teaching career myself I didn't bother with math concepts hardly at all with the children at first. I learned that math and science were my weak areas in my early career. There wasn't much stuff in those centers because I didn't give it the attention it needed and provide interesting things for the children to do there. My defense that I never spoke was this: they are four years old. I only need to be able to count to ten. Although I never said it, it showed up in my attitude. I just very quietly did what I did with those learning centers that had happened with math all those years before with the story problems teachers gave me in school, I skipped them.

As I got better and better at that workplace with the children, I started working on the math and science centers. I became a lover of blocks, counting games and board games. I even designed some math materials for the kids and my boss loved them so much she bought them. I was on my way.

Now as a professor, I have gotten to know a great many of you and studied what makes you successful in learning. I realized that you probably won't become mathematicians. It does no use to make someone learn something they won't use later or perceive they won't. That is a rather strange thing that plays out in a variety of ways. It works kind of like the placebo effect in reverse. As you know, a certain number of folks take a pill that is just sugar and not the real medicine but are told it is, their belief makes them gets better. With this, the need for math is there and they don't believe they need it so they don't see that it is everywhere and in their career in some form regardless of the profession. Except... they do need it.

As a student in a community college, you realize you need to make a good living. You might have kids on the way or they are already here. You might be living with your parents and they do all that planning for you. At any rate, you can get training rather quickly and get connected with those that are successful in your profession and become independent in a new way by making a good living.

I know why you are still reading. You want that degree and math is an obstacle for you. Some of you have to take multiple math classes to get ready for college algebra or an equivalent. It is discouraging. So you put up with working in a job that doesn't require a degree or special training and see all kinds of things you don't like and don't have the power to change it, until the day comes where you just can't stand it anymore. Then what? You take the classes and skip the math leaving it all for the end. Then you either spend hours and hours in the math lab or quit. You got to learn what you wanted to anyway right? Sigh. I hear you. You didn't. And you still can't have the dream job or career without the degree.

So let's get ready for these math classes. Let's take a look at it from a new perspective in the way that you learn best. No, this is not a new way of doing math, I am not that good. It is how I see math and is presented from my point of view with stories, observations, and what I come up with that will be stepping stones to the other side, the math side. Argh! Yeah, I know. You will still be you when you are finished but less frustrated because others might be able to understand you better. You will have some skills you can take to work. You might be able to understand your tutors better who don't get your way of thinking at all and seem to not be able to help you. It has been a life of a math confidant that has been your go-to math person who will do it for you. Trust me I had one for a long while. There is even that frustration of when someone suggests that you use a calculator and you know you don't know which buttons to push and don't say so. You deserve better.

How long will this take? Learning takes time. It depends on how much effort you put into it. Did you ask that for things you love to do? No, because you like it so much you don't even count all the time and effort it takes. I know you just want to be done with it and want to know how long it will take. I get that. Really I do. That is a great question really. It means you are thinking mathematically already because you want to measure it. That is just what I am talking about. You are on the right track only let's put a positive spin on it. You have been missing out on some really cool stuff.

This is all about building up your math awareness. Math is everywhere. I didn't know that for a really long time. Then they assigned me to teach a math and science to prepare child care providers and directors to work with very young children. That was when I understood that math is everywhere. I finally saw it in a new in-depth way with eyes of wonder than dread. I want to pass that bit of wonder on to you. It is so fantastic to have that.

Not convinced yet? It took me a while. This is where I started... I learned a long time ago that the children's book <u>Alice in Wonderland</u> by Lewis Carroll is full of mathematical principles. I had never noticed a one for as long as I have heard the story back then. I was embarrassed and surprised. I had completely missed it. To cover for myself I replied, "No wonder I don't like that story." See, I am with you. Let's go see what we can learn.

Part 2: Getting Cheated is No Fun

Even going to the store has a lot of math awareness possibilities. Besides the obvious prices of things everywhere there are other math awareness principles in a store too. Can I afford it? The sale price percentage off begs the question of how much do I really save? There is a lot of math awareness you need to have to go there and not be cheated in so many ways you might not even see.

Take tax free weekend. The stores are busting at the seams with people. Why? Because they think they are saving big. They are most likely not. My own budget for school supplies for my children came to around two hundred dollars. I bought school supplies, shoes, and an outfit for each of my two children. I split the cost with their Dad who would reimburse me for half. If I went the week before with all the merchandise was already out and on sale, there were no crowds, and lots of selection. They could try things on with no line in the changing room or at the check-out. It was worth it to me to spend the sixteen dollars that I split with their Dad, so it really only cost me eight dollars for peace and quiet, selection, shopping unhurriedly and getting everything we needed.

Another math awareness principle business banks on is that they make you think you can spend to save money. Whenever you spend you will have less money and not more, no matter what. Ever.

Determining a good buy involves homework. You have to know what it is you want and how much it is priced for. Then you look around and see where it sells for less. Quality is involved too as you could get it cheaper and a different brand but it not being made as well. I know you probably do this already.

Another thing to look for is product placement. If it is conveniently located, it might cost more. When I look for a similar item within the aisles at a grocery store, I may pay less somewhere else in the store. It might be a different brand or in a form that might work out just as well.

I never get the store credit card at the register. It is not a buy to save money because what you pay for is hassle. Store credit cards have high interest rates so it is cheaper to pay cash and be done with it. The cashiers can get quite pushy about giving you one of these. I usually level with them saying, "You have to say that don't you?" To which most often they agree. One woman got so pushy one time my next statement was going to a loud "Since when does (name of the store) not accept cash?" Lucky for her, she gave in right before I had planned to say that.

The dollar store is generally not a good buy. Most of the things in it can be bought elsewhere for less than a dollar and sometimes at better quality.

I do not ever buy lottery tickets. Some people who buy them are sure their lucky numbers will work this time. They think it is a skill of the game. It is not. It is a game of chance in which they have no control over. People are more likely to be struck by lightning than to win the lottery. I have seen the math on this.

I am not a member of one of those large clubs to buy things in quantity. It is not a value for me. The quantity is too large for the size of my family. As a single parent, I needed greater variety of products for my family. It is a greater value for me to buy smaller quantities of many things. My budget was too tight to stock up on anything. If I did, other needs did not get met.

What this means for you as a college student. It means that people who can determine these things in their daily lives will use it to measure the quality of many other things. Take this enrollment in community college. It is a great value. For a few thousand dollars you can acquire enough skills to change the direction of your life forever for the positive. They teach you about the value of things and making comparisons of the things in your field. You have the opportunity to be better than average and to make a better living with this training possible. It has a higher value than those without math awareness don't see.

Some people are so bent on not taking math classes that they will gladly pay huge tuition a private school to specialize in something that doesn't require math. Making that high priced education pay-off later is even more difficult because they don't have the one thing they need to do it…you guessed it, math awareness. Those schools know that and guess what? You will get math, they weave the math in where you need it to acquire the skills you need for the profession. They slip it in to the context of the learning making it more palatable because it makes sense in that context.

 Years ago a family member once said, "You get what you pay for." He was a big proponent of paying a little more to get something nice. This rule alone though does not get you a good value. Just because something is more expensive does not mean it is of a greater value. This happens everywhere. Some colleges give "free" I-pads or lap tops when a student is enrolled. Trust me, you have paid for them or it was grant money that will be short lived and you will have to give the technology back at the end of the term. What value will that piece of electronics have for you as a student? Do your instructors know how to use it to bring about excellent learning or is it a google and Wikipedia fact machine or an electronic workbook? Or will you just spend time on social networking sites and gaming instead of study with it. See...I get you.

It is not likely your college will try to cheat you either. They are big about wanting you to succeed. When you look good, they look good. Community college is an excellent value that way.

Textbooks can be super expensive. There is a reason for this. They are so specialized and sell so few of them that they have to keep the price high to make money. These days you can get them online used, or rent them. Some even offer them chapter by chapter and you can buy only the chapters your instructor uses. I have seen people buy one book and share it among themselves too. One student I know got a $130 book for $40. Now that is math sense used well.

Bringing your own food to college will also save you a lot of money.   It is likely your own food will be better for you and cheaper.

Fellow students in a study group you put together is a good value too. Your friendship is the cement that holds the group together. If you did not put the group together consider why they might want you in the group. Consider how much the instructor is going to supervise it too. Groups that are helped to work together do well. That is a good value. It gives you needed practice that is difficult for lots of people to do alone.

Part 3: Managing the Data You Hate To

Something else we do in our lives is manage our equipment by using math awareness. Many of our devices have counters on them so we can interpret how to maintain them, use them, and when to replace them. People without math awareness are completely surprised when their car runs out of gas, their phone data is over the limit, or their bank account is overdrawn. They don't keep track of the data these devices or the messages they are sending to anticipate when the maintenance needs to happen. It is all very predictable and easy to schedule and keep track of for people with math awareness.

Budgeting is similar. We budget money, time and energy. We prioritize what we want to do and how to spend our days, our weeks, our years, our lives. Some of us feel a sense of control over it, others a sense of fate and luck. Our point of view dictates how we manage all the resources we are given. Some live by a schedule or plan and others live in panic mode where most everything is an emergency. Some feel such a lack of control over their lives they do nothing and wonder why things don't happen for them. Truth is, they don't invest enough of themselves in planning to make it happen. They don't see all the small steps that add up to big things. They are waiting for the big moments that never seem to come. They have poor math awareness in how to invest in themselves because they don't know how to budget. They expect to be rescued when the resources run out. Sometimes they do have people who will do that for them and live their entire lives that way.

I know someone that burned up a perfectly fine car simply because he did not change the oil. The block cracked in the engine and it was a major repair that left the car pretty much totaled. He was so stunned. He did not leave it to anyone else again. Amazingly, he got rescued and got help to get a new car. This continued rescuing has got him in a kind of trouble that he did not anticipate. He gave away his power because without math awareness he didn't know what he could afford in regards to cell phone service, a computer, or other math awareness decisions he could have made himself. He became connected to people who would do this for him and then they made demands back of him which he felt obligated to fulfill. He is in quite a mess now. I was with him when his phone rang and watched his face change to fear as a call came in from this person that rescues him. This person "has him by the nuts" as they say and he can't find a way out. Without math awareness he doesn't know that he is closer to freedom from this situation than he thinks.

In regards to energy, I was such a mess about this too for a good long time. I gave all my energy to people and eventually burned out and was depleted from all the so called obligations I could not say no too. When I was depleted and needed help, nobody came. I was shocked. I was hurt. I was resentful. I grew bitter for a while. I discovered most everyone in my life at that time was not supportive of me and didn't really give a care because I could no longer meet their needs. They went elsewhere. The others who did care, I didn't tell. I was too embarrassed. All of this happened because I had no math awareness over my own energy. I didn't actually know it could run out and it did. I learned I had to replenish myself and bit by bit learned to say no. I had to build myself back up and find new people.

Budgeting time or the lack there of is all over. Some people are so bad at it things like Christmas shopping and preparation can be completely overwhelming. This is a cultural phenomenon as many people believe that happiness can be bought. They fight with others in the store for coveted items, cut people off in traffic for prime parking spots and get short tempered when the check-out line is long. They max out credit cards and spread the financial mess over several months all for this one day of happiness that never quite measures up. They lose all math awareness in terms of their resources of time, money, and energy. They believe for a time that crazy Aunt Kate and drunk Uncle Joe are going to be different for a day. They want to believe in magic and that things could be better. For lots of families this cannot happen.

I am one of those who decided years ago that I was not going to be part of it. I didn't get caught up with the frenzy. I shopped early. So early in fact that when my kids gave me their lists of what they wanted, I told them I had shopped already. They were so surprised and delighted that I would find something just for them that would be a surprise. That is the magic the holiday needs as I see it.

People in my life ask me if I am done shopping as the days get closer every year. I say yes. They often tell me of their lack of time. I finally said it like this one year, "Christmas comes on the same day every year. It is not like Easter that they move around." I implied that they didn't plan for something that has a countdown for goodness sake. They didn't like that.

I worked for an older woman who knew better than that. She was especially good at providing a good value and rewarding people who gave her more that she expected in terms of their work. She had us all over to her house for a staff Christmas Party and told a story of how she and her new husband, then deceased, spent their first Christmas together. They were so broke they bought a very small Christmas tree but could not afford any ornaments. So they bought a paper nutcracker and attached it at the top and pasted red hot candies all over the tree. That was it. It was so heart-warming to hear her talk about it and we could hear how much she missed him after passing away a few years prior to that time. There was a hush over the room and we looked at her 8 foot tree all covered in nothing but nutcracker ornaments they had collected over their many years together and people had given her since. The few dollars they had, had given them a value that surpassed what they could see. Many years later there it was for all to see because of math awareness.

<u>What this means for you as a college student</u>

As a student, you have to monitor your own energy, time, and data. How you do this has a huge effect on your success.

One of the most common things I see is students who take too many classes in your first semester. It is easy to get all excited, get overwhelmed and then quit all because you didn't plan your time, energy, and data. You didn't give your advisor the information they needed to help you make good choices.

At first look, a college schedule looks like nothing. There are lots of gaps in it. This is only class time. It doesn't take in account study time, lab, or homework completion. When they ask you if you are a full time student, they are not kidding. Five classes does take 40 hours a week to complete if you want to do well in them.

The course load is something to consider too. I know someone who enrolled in a community college and had trouble reading. They enrolled him in 4 classes all that had massive amounts of reading in them like history and literature. He was unknowingly set up to fail. He never went back. With math awareness and knowing that these classes require a lot of reading he could have spoken up for himself and given himself variety. A PE requirement, and for him, a music class would have lightened the load considerably with only 2 classes with reading. He didn't know how to divide up his time for socializing and prioritizing either. I don't know if he ever will go back to college.

Most people in community college need to work either part time or full time. These hours are best completed <u>after</u> you study. The best employers you can have while in college is one that will give you a flexible work schedule.

I know right away which students are the ones spending all their energy. They are the ones that are continually sick with the flu and sinus trouble. They miss a lot of class because they realize they are in over their heads. Being hugely stressed out can make them instantly sick. It has happened to me on several occasions. It is okay to take college slowly at first and then build as your "muscles" get stronger.

Family and social time matters during college. A lot of people pack their schedule so tight that they forget to leave time for fun. You will most likely meet new people you want to hang out with. You will want to have time with your family. All of this is super important. You want to have those "when I was in college" stories to tell your kids about how fun it was.

Then there are those things I mentioned earlier about everything else in your life that needs monitoring too. It is a lot to do. With math awareness you can get on your computer and check it all in a few minutes.

◊◊◊

Part 4: Keeping Track of Your Own Success

As I mentioned earlier, there are people that measure things in different ways: some take control over it and some believe in luck or fate, that others are just lucky or have natural talent. There is a third group too that have so given up on learning that when something like math comes along they withdraw to the degree they are barely present in their heads. These both have names you can look up and study "locus of control" and "learned helplessness". There is a lot of research on both. These things puzzle great teachers who want to see you succeed. It makes us as professors want to visit happy hour on the way home from work or take a trip to the Bahamas at the end of the semester to get over our own disappointment. We can see the promise in your eyes and talents. We see the many road blocks in your way. You can do something about this by improving your math awareness.

I had a student in my class who had been working on her degree for many years. She has a passion for what she wanted to do and a maturity that astounds. She worked hard the entire semester. She revised when I offered it to her. She missed classes and once on a test day and had to take it later in the testing center. She did not keep track of her grades at all on grade record sheet I had in the syllabus. She expressed to me the last week of class she hoped she passes the class. She got a B. What I would have given to see her astonished face when she looked up her grade. She had very little math awareness. She would have gained more confidence if she had only the skills I mentioned earlier and kept track of it. With math sense, it would have been an easy A for her but she did not know it. She spent the entire semester thinking she was failing and worrying.

I also have had a few students who are all math awareness. They plan, they schedule, they revise, they keep track of their grade, and they succeed. Funny thing is, not all of them have confidence in their abilities. Confidence is easily observable, it separates the excellent from the astounding. Believing in yourself is not all math sense. It is definitely tied to emotions and experiences too. Truthfully, I sat in my own graduation for my master's degree wondering why I was not prouder of myself. That took a lot of work that is another book <u>Taking Care of Susan</u> that I wrote. It is the story of how I began to take control over my life after my master's with all the know-how I had.

The lack of math awareness shows its ugly head in many ways in regards to success. I had a student that told me college was more work than she thought and told me what her four year old child asked her, "Don't you like us anymore Mommy?" I knew when she said that, she was gone. I gave her a pep talk anyway about how this will change the quality of life for her and her children and it is only for a little while. She looked at me saying nothing and she left. I never saw her again. Funny thing is, my own children said something similar to me as I was working on my master's degree. This is what I answered, "Do you like ice cream?" They said, "Yes." I said, "When I finish school we don't have to get ice cream cones anymore, we can get sundaes." I smiled and I nodded. They lit up. I took them to graduation and let them know it was over when it was. It only took seven semesters. Today, as adults, they can barely recall me going to school. They remember eating from the snack machine and walking through the indoor garden in the atrium. I knew the value of what I was doing and was able to convey it to them. I was able to measure my own success.

Many of us are intuitive and measure success by how we feel about things. Good as we are about it, sometimes it takes math awareness to give us the whole picture. We collect compliments and testimonies, all which are important but not enough. There are two things to look for: to measure what is happening and to identify things that are missing. Before the math people reading this get too excited, it really is important that both things are happening: intuition and hunches go with the numbers to make a whole story. We cannot have one without the other.

<u>What this means for you as a college student</u>

If your instructor gives you a form for keeping track of your grade, use it. Record everything on it as you go. Watch how much everything is worth. Some projects are worth a lot of points and some are not, this could determine how much time and energy you put into it. Some students put too much energy into things not worth a lot because they like it or don't see the importance of it in their whole grade.

Not all professors give grades periodically as the semester progresses like in high school where grades are due every 3 weeks. It is not a requirement, at this time, in college. In some courses the work builds into a large project at the end. They can do that.

What is not acceptable is not getting your work back in a timely manner. Your professor should have designed the course for you to build your knowledge over time with feedback of some kind on all of it as you go. That is <u>their</u> math awareness problem *if* you get it in on time and weeks and weeks pass without getting it back.

One last point to consider is…does it really matter if you get your work done late? In a word, yes. They have set it up for you to get the most learning out of it as possible. By changing the pacing of the class yourself, you won't get it all done. You will simply run out of time. This is particularly bad if you are in a program that has a state certification test at the end. Think about it, you, the amateur, are trying to out-guess the professional who has experience in the field. Not good. Even if there are loop holes, don't go there. You need to learn all you can to have this college degree.

Part 5: What's Missing?

Don't ya hate it when you order something in the mail, get the package and a screw is missing? You got the whole thing together that clearly won't fit in the box now to return it and then you are in a jam. Had you looked it all over before you began, it would have been clear. Who does that? Practically nobody because most of the time the pieces are all there in everything you order most all the time. It simply isn't an issue. Having math awareness means that you need to take into account what is not there sometimes.

People who commit fraud count on it. They slip away money or objects that nobody knows is missing because the others don't see it. The people they work for are not stupid or anything, they just don't notice. Then it adds up to a lot and they either run off, which signals a problem, and they go to jail.

Part of any job is sometimes to notice what is missing. It is also where innovation happens sometimes. One of the conditions for innovation is to fill a need people don't know they have. For example, when food was first preserved by canning, people did not know how to open the cans easily. The can opener was invented later. Seems strange to us today, but that is how it happened. Someone solved a problem for what was missing.

Having math awareness means you know to look for what is not there and should be. It can happen on paper, in the store room, in the relationships with people. There are things missing everywhere.

<u>What this means for you as a college student</u>

With math awareness you will be one of the students who seeks out other resources to help you understand what it is you are trying to learn. While others have excuses not to, you gladly dive in because you know that you can seek help of the professionals at the college to give you excellent resources and not google and Wikipedia.

You will have questions pertaining to your field that comes up in the learning process that are "off topic" you can ask your instructor later. These questions could be an honors or extra project. No, you are not seeking extra points but to increase your learning. You will make yourself unique and marketable to future employers that way.

Why would you want to do this? Because once you get into your work, it is going to feel like a desert at some point, completely empty of all these people and resources. Although you will have professional development, many places choose to do their own. This can lead to stagnation in the workplace because the same information goes round and round.

Should you do get sent to an outside training, leave your group and go sit with other people and start conversations. So many people go in groups and do the round and round thing in a far-away place. It does come off at weird to them that you are everywhere else but with them, but you are seeking out new ideas to make it fresh for yourself. You want to stay connected to your work. Sometimes it is the opportunity to discover something that was missing to a problem you struggle with at work that many other people have.

Part 6: Patterns are Every Where

Patterns are in the weather we experience, the flowers and foliage of your locale. They are in the growth patterns of your children and how you spend your free time. They are even there in the things you "like" on social media. People tend to post the same things all the time with their favorite topics. I invited a friend over for dinner once. As a thank you she brought flowers. I was so delighted and surprised. How did she know I love flowers so much? Social media and math awareness that's all. She saw the pattern.

They say that people know each other better these days because of social media. They know so much more before they marry because they know what each other likes. Some say that it took our grandparents or great grandparents 25 years for them to know that much about each other.

For many people who do not have math awareness, math problems seem like a group of random numbers. Sometimes I am still right there with you in that one, once in a while. Trust me, there are patterns in the math problems. You can learn to see them.

There will be patterns in your work habits and how you will deal with situations that even you don't see. Your boss will see these plain as day and help you to improve these things.

In your new work at the beginning, everything will be a surprise without work experience in that field you don't know yet. After a time, you will begin to see the patterns. People tend to ask the same questions and habits. You get better and better at fielding and recognizing them. Next thing you know, you a success at what you do because you anticipate what the job needs because of the patterns you recognize.

What this means for you as a college student

As student you need to know that your college transcript is a document of your education. Without math awareness, it is a list of grades and courses. With math awareness, it is a story. It tells when you began, what your weak courses are, what your strengths are, if you change your mind a lot about what you want to do. It could be a key to career choices and how successful you might have in a job, all based on the story it tells.

A lot of students have a set group of patterns in how they go about learning. For the successful student, nothing needs changing. It is interesting that those who are unsuccessful also have patterns to their study that are not working. I have seen people continue with these unsuccessful patterns as they repeat the same class in the same way and expect a different outcome. Suddenly, the instructor becomes the devil when it is the lack of a change in their patterns that needed to happen. A good instructor will suggest other strategies, if they see the pattern. Don't count on it. If you do something the same way, it is highly likely you are going to get the same result.

Changing your own patterns is hard work. There are people on campus that can suggest things to you to try but only you know you spent too much time online or partying or sitting discouraged. It involves stepping out of your own comfort zone. This can be uncomfortable for you as you abandon old habits for new ones. It depends on how badly you want to succeed.

Part 7: Practice

If you turn on the TV, you are going to find lots of success stories. These stories not only amaze and entertain but they cause some people to think unrealistically about their own success. What appears to be a chance moment with a microphone that launches a music career is actually the end result of a lot of practice.

I have seen people choose majors or focus their major based on something wonderful they saw. It is almost a kind of magic to see someone so accomplished at something that it makes it look easy. Be careful with that, it is not easy no matter how it looks. Use it as a source of inspiration but don't for a moment think that what you saw is easy for you to learn too.

Anything we want to learn is going to take practice. It is going to go better and faster with a teacher. They know that they are going to need many interactions with you to help you learn. They invite you to the process and adventure of it. They like to see understanding happen. Trust me, it is a different look on the person's face when they do understand than when they don't.

You could go it on your own if you are stubborn. Many people are self-taught and do quite well at things. They even make a nice living for themselves. This involves lots of trial and error in the school of hard knocks for some. Some really famous people even quit college to do their dream. It hits the news. Of course it does, it is extraordinary. What they don't tell you is that the famous person discovered they already had the skills and practice they needed to do their thing, lots of practice. Typically, the average person that quits college does not have skills and practice. That is why it doesn't go so well for them. Going to college is shorter than trying to figure it out for yourself, it just is, for most people.

If this dream of yours is important and you leave college to try something else, you will be back. A real dream has a way of tapping you on the shoulder over and over again until it grabs you by the throat and pins you up against the wall insisting you do it. Cool visual huh, if only it is that clear. For most people it is more like an urging that pesters for years leaving you disappointed and regretful later. Consider digging in now. Go to the math class with an open mind. People work long and hard to put together the program you are in. They have gone before you and know what things you need to know to be a success in the field. Practice is important.

I made that mistake too. I decided I didn't like my child development class and thought I was so lucky about not having to take a computer class that was going to be required for graduation for the class behind me. As you can guess, these things are the center of my work now. I had to practice on the job and make mistakes.

What this means for you as a college student

Most students new to college seriously underestimate how much practice they will need with the material in order to learn. It takes at least seven repetitions of the same material in different ways in order to "get it". It is highly unlikely you will get all seven in class or in the things your instructor has you do. You are going to have to come up with some of those practice sessions yourself. As a less successful student way back, I studied what I knew over and over again hoping I would get all those right on the test. I got C's and D's that way. Later, with more confidence about what I knew, I set what I knew aside and focused my study on what I didn't know. My results were A's and B's.

Remember the potato chip story I discussed earlier in this book? My instructor didn't give it to me. I found it in the bookstore on video tape when I went to look for some kind of study aid. I changed my own pattern with practice. Back then, there was no internet or you tube. As I see it, you haven't got an excuse.

◊◊◊

Part 8: What Math Concepts You Need to Learn for Your Career is Not Your Responsibility

It really isn't. Your instructor in your major knows how you will use everything it your degree plan. If you don't know, ask. If they don't know, it is their homework and not yours. You deserve an answer to a really great question. It is not the responsibility of your math teacher who deals with all majors. It needs to make sense to you why they are requiring it and how it matters in your field. So when you are working on learning it, you can remind yourself why. It is easy to get lost in a swirl of assignments and tests. It is easy to lose your direction because it goes so fast. Fast? Since when is four years or two years fast? One day you will look back and agree.

Some of it is for a general education. It is not high school all over again. It is critical thinking skills and other skills that is going to make you well a rounded professional. You will not be asked to learn a general education type program any more in college if you go on for graduate degrees. Your future instructors will assume you know and will draw upon what they think you learned. You don't want to get caught in a situation where you are taken completely by surprise. Learn it all and learn it well, including the math.

What this means for you as a college student

Going to college to prepare for a career is like packing for a trip to somewhere you have never been before. It is hard to anticipate what you need. When I lived in Illinois we went to Houston for Christmas. I brought black velvet pants as part of an outfit to wear on Christmas as dress clothes. When I got to Houston I realized nobody wears black velvet. Worse than that, it was 70 degrees that day. So not only was I out of step, but hot too. College is like that. At first it can be full of surprises and wondering how you belong.

Attitude can hold you back on this. How you approach what it is you need to learn, matters. It matters a lot. Students who come open minded are easier to work with than those who can't be bothered with what they see as "unnecessary". They are easier for the instructor and get along better with the other students. The wrong attitude can trash your opportunity to get a letter of recommendation at the end of your program, just sayin'. You choose. It's your future.

Part 9: Thinking Styles

Because it is "fun" or "feels right" is not an answer a boss wants to hear, trust me I have gone there. They want time lines and plans. They want budgets and projected outcomes. Taking math classes helps you to develop such a plan in your work someday.

Personally, I still find this very frustrating to pull some plan out of the air and put it down on paper. It doesn't happen that way for me. There are two kinds of logical thought: deductive reasoning and inductive reasoning. Deductive reasoning is like the thought that goes into a science experiment. There is a goal in mind that people work toward. Inductive reasoning involves a lot of collection of data and then pulling it all together to a conclusion built on that evidence. Personally, it takes me twice as long to do the deductive. My thought process works best with lots of "pieces" to work with. If I have stories and some "things", like math manipulatives, I am there. Maybe you are there too.

When I taught young children math, I did just that. I gave them things to touch that had kid appeal and then we would take out strips of paper and write down the number sentence for what we did. There had to be a connection between what we did and how it looks written down. May you missed out on that part and they went straight to numbers. I used both thinking styles with the children.

College math instructors need to think about that. If yours is not, than it is time to figure out your learning style and make it happen for yourself. You are an adult. You can find out how you learn best and then adapt it for yourself. If you can't do that then go ask in the math lab. Tell them your learning style. Ask them to use that style to help you. They should be able to help you with that. If they can't, someone on campus can. Go knocking until you get it.

When you ask, have an open mind for a conversation with this person. Let them know you want to succeed. Let them know this is part of your dream. Listen carefully as you get your answer. Be ready to ask more questions about their answer. Whatever you do, do not stomp out of there and slam the door. Math has been a point of frustration for you for a long time. Go home and think about what they said and go back in a few days to ask more questions. Keep going with this until you understand. You might end up with a math partner who will help you learn and not do it for you.

I did that at the community college with my college math instructor. I went every time before class. I got more and more comfortable about asking. He learned to expect me. It was a great experience.

If you find you cannot "get it". You have many possibilities on you tube. Two of them are Khan Academy and Yay Math. They explain it differently than perhaps your math teacher and/or give you practice. Some math instructors do use videos such videos in their teaching.

What this means for you as a college student

Out in the world and society there is the idea that the customer is always right. I heard some insight from a businessman who explained this. He said that the customer is not always right but the customer is the customer. They are their source their livelihood comes from and to treat them well is a good idea for the reputation of the business.

As a customer, you want your questions to have math awareness. People without math awareness sound terrible in what they want as a customer and make impossible demands. There is this story circulating the internet of this woman who called about the signage along a particular road. It was a deer crossing sign and she had hit a deer and wanted it moved. She was sure if they moved the sign, the deer would not be there. You should be laughing. If you don't understand this, run don't walk, to math class. You need to develop the thinking skills to be able to understand this story.

As a student, you want to be one who asks good questions. You don't want to be the deer lady. It is okay to ask an instructor questions outside of class until you get your math sense developed. When your instructor tells you that question should be asked in class so others can benefit from it, then you are doing really well and should ask in class.

◊◊◊

Part 10: Loosen Up

TED Talks provide lots of unconventional ideas about how to use math in many different ways. They apply it creatively. They "play" with math. They use it to find love, use it in slam poetry, they see it in coral formations. Watch some of these videos and see the fun you are missing out on. Wouldn't it be nice to have enough math background to be able to laugh with the audience?

<u>What this means for you as a college student</u>

You will learn more if you are relaxed, plain and simple as that. I have tested it myself. I added skits and silly piggy back songs having the students put the course content to a familiar tune. It was hilarious. Students loved it. I loved it. So much more learning happened in the classes that laughed a lot. Self-esteem grew in them too as their confidence grew.

◊◊◊

Recap

So this is what your life will continue to look like without math awareness:

You will be anxious when you encounter math.

You will have a go-to math person you depend on that may or may not be available when you need it.

You will most likely be cheated on something.

You might have difficulty determining if something is a good value.

You will continue to be surprised when your data runs out, your energy runs out and you run out of time.

You might continue to leave success to fate and luck.

You won't see what is missing when it really counts.

You won't see that patterns and math are everywhere.

You don't see the value of practice to improve your situation.

You might blame others for your lack of knowledge.

You won't understand how you learn.

You won't get your dream that only graduating from college can get you.

What you could gain by gaining math awareness:

Confidence in your own abilities.

Increased self-esteem.

Informed and savvy decision making.

Control over things in life you do have control over.

You will have accomplished something that is very hard for you. Such accomplishment leads to more risk taking for future successes.

You won't depend on others for things you shouldn't and feel a sense of healthy power.

You will have a degree or certification that will be an opportunity to walk in the door at a chance to a new life beyond what you have now.

Ending Thoughts

So now that I am finished with this book, I have a surprise for you. I wove math concepts into this book. Some of you probably saw it. I used plain language as much as possible to convince you to give it another chance. Your dream is worth it, dig in and dig deep.

Here are the math concepts:

Part 2: value which involves addition, subtraction, multiplication, division, percent, making comparisons, estimating

Part 3: budgeting, addition, subtraction, ordering, comparing

Part 4: measurement, statistics

Part 5: algebra

Part 6: pattern recognition

Another surprise I have for you in this book is it is not all math. It is a reflective and intuitive process too that involves self-reflection and personal growth. I wove these in too.

Here are the personal development skills while learning math:

Telling the truth

Increased intuition

Balancing interdependence with independence

Acceptance for what we cannot control

Responsibility for what we can control

Delaying gratification

Tenacity

Focus

Timeliness

Self-reflection

Ability to laugh about math related fun

It has been a pleasure working with you. I do not believe the world needs more STEM (science, technology, engineering and math) or more STEM majors. I believe the power than causes us all to breathe knows best on who should do what and how many will be needed. It is super important to follow your bliss no matter what frustrations and joys that contains. There is no easy path, no easy majors.

What I do know for sure is that we do need all the tools we can find to become accomplished in something. We need math, science, logical thought, literacy in all forms, abilities involving creating pictures and diagrams, a love of people, an opportunity to be reflective, awareness of history and the present, a love of the arts in all forms, beauty and willingness to give back. Community college fills your toolbox to be ready for anything your work and life might ask of you.

Since I really don't look or act much like people's idea of a professor, I get to hear people's views about college in the unlikeliest places where people will strike up conversations, like waiting rooms, with people they don't know. One of the recurring themes is the desire to go to or to finish college. There is regret in their voices of a missed opportunity. Their situation or their choices as a younger person disappoint them. They would be the first ones to tell you to continue.

Someone believes in you. They have provided the money for you to go to college and have an opportunity to something adventurous and expanding through learning. Whether it is your Mom or a governmental grant providing the money, seize this opportunity. You are not too old.(Our oldest graduate at the college I work at is 83.) Community college is for everyone that wants it.

Back to the conversations I have been part of...no one has ever told me that they think college is a waste if they already have a degree, ever. And please, whatever you do, don't stop because of a math class. Whether you know it now or not, you are worth it. You really are. You can exceed the expectations of those who do not see the promise in you. People have the amazing ability to grow and change. Put those naysayers on the floor with surprise or scratching their head with disbelief. I have seen it myself in others and for myself. It is super cool when that happens.

Once in a while, a student will come back to visit at the college where I work after achieving their dream. I find I have a lot of trouble recognizing the confident person before me that was so unsure before. I love it when that happens. Let it be you.

◊◊◊

A Math Theory and Sparkles

I was at a bookstore and picked up a book of math theories. It was a small book and I thought I would give it a look because it didn't look that much like and ordinary math book. Here is what happened. I randomly opened it to a page on Catastrophe Theory, I kid you not. I read a little and about burst out laughing. What it said was that things can go along as normal and predictable and then one thing happens and it goes another whole direction. There was even a detailed graph indicating how it happens with the line shooting off in another direction. I loved that.

What this means for you, for the right brained dreamer is this: math can explain those sparkles that sent you to college to pursue your dream and those feelings that go with it, it just makes them nervous that the unpredictable happens. Taking the math classes are going to help you get there, down the road, to that beautiful thing you imagine. It really will. Even if it is not a big part, you still need it.

◊◊◊

FAQ's:

Q: So where do I sign up for your math class?

A: There is no math class. I am not a math teacher. There are lots of good ones out there. I wanted you to be a success, to work your way to your dream. It's the dream that matters to me most, not the math, and removing as many stumbling blocks as I can for people. Apparently, this pep talk worked.

Q: Where did the idea for this book come from?

A: I saw way too many people not graduating because they didn't want to take math. I think it is a necessary step for some to back up and take a look at how this decision affects their life.

Q: If you are not a math teacher, what do you do?

A: I train people to be teachers. In my off time I am a writer. I self-publish my own books. I have a writing studio in a public place to prove to myself how serious I am about it. My dream is to one day have a creative life with a creative job.

Afterward: an irresistible ending

So, a few weeks ago I decided to retire from teaching after 34 years in the field, 29 in the classroom, and 20 at my last workplace. I am pursuing my creative dream of writing and other creative pursuits. You know what? I do need math for this dream and I did know that I would. I put everything in place so that even the most logical of naysayers, who don't want to hear of sparkles, could not argue. I wrote 22 books since 1980. I sold 315 copies since 1989 of 9 different titles, selling at least one copy of each title. My retirement check is my financial safety net. No one can even call me a starving artist because thanks to my teaching career, I am assured of a lifetime of eating, medical insurance, and other necessities.

It has greatly surprised a number of people in my life. Okay, it was more like shock and awe. It has been an inspiration to a few too that there were possibilities to retirement for them too if they planned. I too thought retirement meant to go to the senior center and play Bingo. No, it can be a vital time of following your bliss, whatever that might be. And for the record, I am 56 years old. Even too young for the senior discount at most restaurants because I started early and kept going. Math awareness pays. Get after it. Something beautiful is waiting for you.

www.ingramcontent.com/pod-product-compliance
Lightning Source LLC
Chambersburg PA
CBHW071006180526
45168CB00003B/1315